Mommy, Where Do Butterflies Go?

Leta Laugle

Blue Ink Media Solutions

Mommy, Where Do Butterflies Go?

Printed in the United States of America
ISBN 978-1-967279-06-7 (sc)
ISBN 978-1-967279-07-4 (e)

2025.02.11

This book is printed on acid-free paper.

Blue Ink Media Solutions
1111B S Governors Ave
STE 7582 Dover,
DE 19904

www.blueinkmediasolutions.com

DEDICATION

I would like to dedicate this book to my mother, Sally. I was visited by the rare, yellow swallowtail butterfly the day my mother passed and every day thereafter for many weeks. It brought me great comfort knowing she was there with me through my grief and my healing. I hope this book brings you similar comfort, as you are reminded, or realize, that we never truly lose those we love the most. This book is a dedication to all of those we love who have gone before us and those we cherish.

Filled with wonder at the daily miracles that surround her, a young girl loves watching the progression of caterpillars into full-grown butterflies. She takes great delight in watching them take flight, yet after one such happy occasion, she returns home to learn her grandmother has passed away. Sadness overtakes her while remembering all the wonderful times they had spent together. What helps in her grief are frequent visits by butterflies, which lead the young girl to ask her mother where they go after leaving her. Her mother answers that God sends comfort when needed and that the young girl must remain faithful and keep her eyes and heart wide open. These words comfort her, and the butterflies become a way to stay connected with the memory of her grandmother.

Explaining the loss of loved ones to a child is always a difficult conversation, but this fully illustrated book does a graceful job of doing exactly that. It relates the progression of life to the transformation of caterpillars into butterflies, emphasizing the divine nature behind such cycles. An inherent beauty gets woven into each step, even when the butterflies depart for distant places. As such, this story is a good choice to read to young children who might have experienced the passing of someone dear to them for the first time and who need confirmation that those loved ones are never completely gone. Inspired by events in the author's life with visitations by yellow swallowtail butterflies, it certainly fulfills the intention of offering comfort amid grief.

Joyce Jacobo
US Review of Books

I like to take walks on a summer
day to see the flowers bloom
and watch the clouds float by.
I particularly like to watch
a small, fuzzy, caterpillar
inch along on the ground.

I have watched a caterpillar build its amazing little cocoon house and hang upside down on a limb of a twig. After many days, the caterpillar decides to wake up, but it will no longer be a caterpillar, at least that is what Mommy and Daddy told me because God would show me a miracle very soon. So, I waited and watched.

Very soon, the most beautiful, big yellow butterfly came out of that tiny little cocoon. It took a very long time for the butterfly to spread her wings but once she did, she flapped and flapped until she took flight.

She slowly went from flower to flower, staying for a second or two it seemed to eat a little bit, then she came and rested on my finger, almost to say "Hello"! Oh, how I love butterflies! They are such happy creatures! I wonder where they go?

That same day when I arrived
back home, my parents told
me that my grandmother had
gone to heaven. All that happy
feeling I had with the butterfly
in the garden had vanished
and now I was so very sad.

I would not see my grandmother anymore. She was so special to me as we made cookies together, sang songs, and took long walks outside.

Mommy, Daddy, and I went to church for Grandma's "Celebration of her Life". It was a sad time.

Then a butterfly came to be
with me which made me happy.

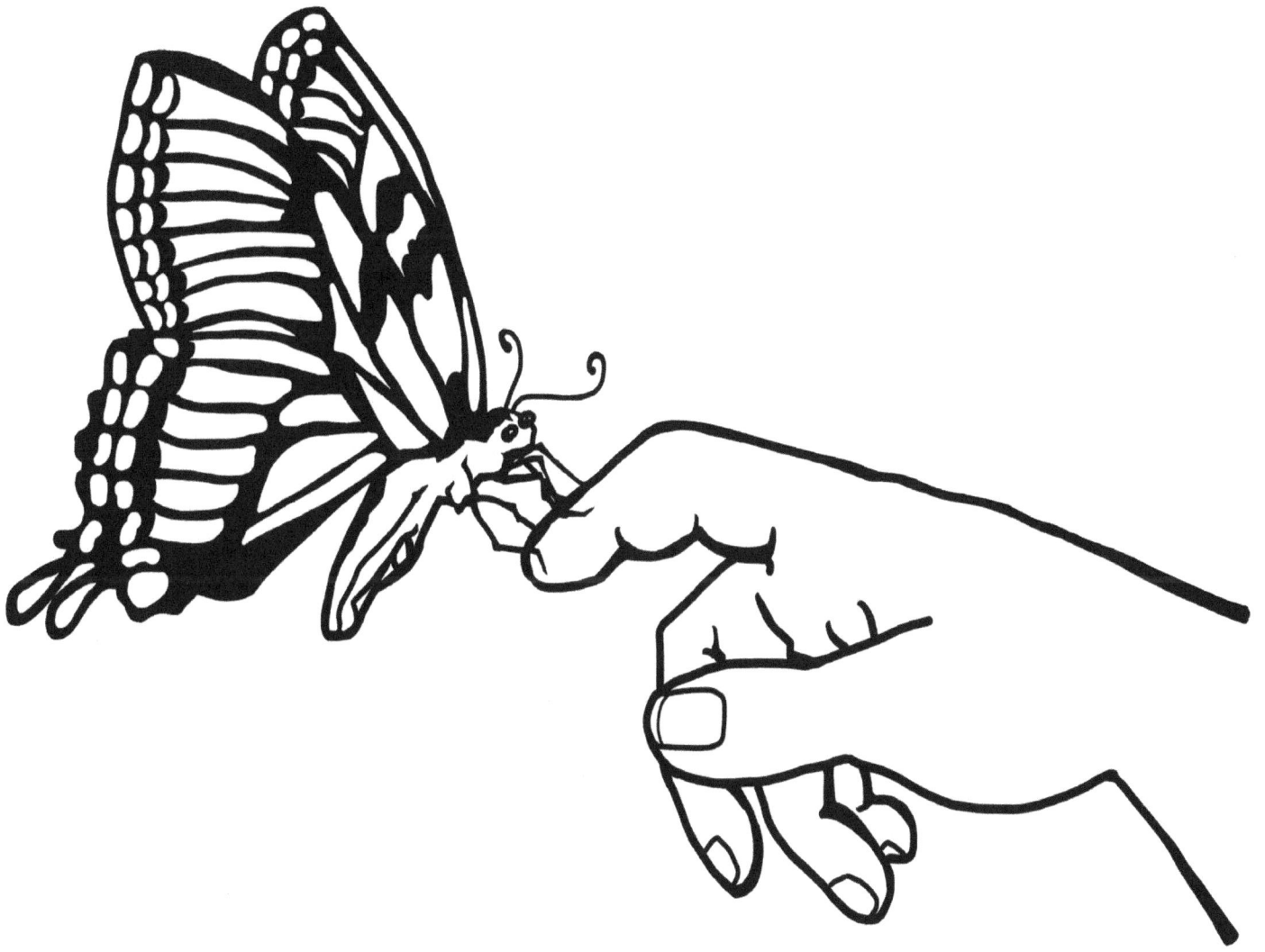

The next day a butterfly
came to my window, again
it made me smile!

The following day I saw
another butterfly.

Finally, I ask "Mommy, where do the Butterflies go? Do Butterflies go to Heaven, too?" Mommy's reply, "God is all around us. He sends us comfort whenever we need it, like butterflies. All we have to do is trust, believe, and pray."

To My Illustrator, Erika Cooperman—
Thank you so much for all of your
hard work, professionalism, creativity,
and most of all, beautiful artwork.
Without you, this book would not
have been completed. Thank you.

To My Loving Husband, Jeff—
Thank you for all of your
love and support. LUTT.